T0281606

Out of the Comfort Zone:

New Ways to Teach, Learn, and Assess Essential Professional Skills

An Advancement in Educational Innovation

Synthesis Lectures on Technology, Management, and Entrepreneurship

Editor
Henry E. Riggs, *Founding President and Trustee Emeritus, Keck Graduate Institute*

Out of the Comfort Zone: New Ways to Teach, Learn, and Assess Essential Professional Skills — *An Advancement in Educational Innovation*

Lisbeth Borbye

ISBN: 978-3-031-01444-4 paperback
ISBN: 978-3-031-02572-3 ebook

DOI 10.1007/978-3-031-02572-3

A Publication in the Springer series
SYNTHESIS LECTURES ON TECHNOLOGY, MANAGEMENT, AND ENTREPRENEURSHIP

Lecture #2
Series Editor: Henry E. Riggs, *Founding President and Trustee Emeritus, Keck Graduate Institute*
Series ISSN
Synthesis Lectures on Technology, Management, and Entrepreneurship
Print 1933-978X   Electronic 1933-9798

Disclaimer: *The content of this book is based on the author's personal knowledge and experience. The author disclaims any liability or loss in connection with use or distribution of the information herein. Use of the information is at own risk. All anecdotes herein are fictional although inspired by events, which have occurred as a result of using the described teaching method. The contents of this publication do not constitute legal or professional advice. Readers should not act or rely on any information in this book without first seeking the advice of relevant experts.*

Out of the Comfort Zone:
New Ways to Teach, Learn, and Assess
Essential Professional Skills

An Advancement in Educational Innovation

Lisbeth Borbye
North Carolina State University

SYNTHESIS LECTURES ON TECHNOLOGY, MANAGEMENT, AND
ENTREPRENEURSHIP #2

ABSTRACT

Success in careers outside the university setting depends on an individual's capacity to master professional skills and respond appropriately to dynamic situations with flexibility, adaptation, and innovative thinking. This book describes a simple, common sense method of how to include professional skills training in any curricula without compromising academic rigor. It relies on introduction of unanticipated yet manageable crises simulating scenarios commonly experienced in the workplace. The method promises to inspire both students and their teachers to conquer new territory outside their comfort zones. Examples include how to respond to a demand for innovation and teamwork, a lay-off, a re-organization, or switching jobs and projects. These situations are bound to occur for most people and in most jobs they often create stress and, perhaps, despair. Preparing and practicing a mindful and healthy response is beneficial, and now this process can be performed in the classroom, while it serves as a platform for character building prior to unexpected real-life events.

Key Features:

- Description of the importance of, incentives for, and rewards of exiting the comfort zone.

- Principles for teaching and learning professional skills.

- Student anecdotes and reflection.

- Rubric entries and assessment of learning.

KEYWORDS

professional skills, teaching professional skills, professional master's education, need-based curricula, innovation

Contents

Preface

Acquiring essential professional skills and having a willingness to learn are necessary to excel in many important areas such as interpersonal interaction, creative thinking and innovation. Inspiring "aha-moments" relevant to an individual's life experience are important for cultivation of a continued appetite to learn. Today, many students at our schools and universities are bored by the traditional teaching methods and have become complacent about learning. Once they have deconstructed "the system" of how to get a good grade this may be all that matters to them and all they accomplish in terms of learning. This tendency has a negative effect on the students' ability to think "outside the box" and to adjust to new situations. Long term consequences include: a less effective educational system, a decrease in intellectual capacity, and lack of ability to cope with change.

I have developed a new teaching method, which focuses on essential professional skills acquisition to counter this trend. The primary principle is conditioning to embrace the richness and wisdom derived from overcoming obstacles and understanding opposition. The goal is to bring students (and professors) out of their comfort zone, hence the name "Out-Of-the-Comfort-Zone (OOC) teaching and learning." In brief, this method requires flexibility of the format for both teaching and learning. It employs continuous alteration of the conditions under which students work on their assignments. It is an interactive, assertive and a "real-life-simulation" teaching style with high demands of responding "on-your-feet" and essential professional skills such as flexibility, discipline and teamwork. The method fosters adaptation, alertness, emotional intelligence, and innovation and can be applied to all fields and at all educational levels ranging from high schools to universities. For reasons of simplicity, I here refer to universities, professors and instructors, but due to the versatility of the described method, these can be interchanged with schools and teachers. Industry, non-profit organizations, and government are here defined as professional environments outside the universities. I often use the industry environments or companies as examples for all of these.

This book is a result of the insight I have gained using this type of teaching at North Carolina State University. A guide to the teaching method and examples of student transformation are provided, as well as suggestions for how to measure the outcome of this type of learning. It is my hope that many educators and students will feel inspired and understand the interconnectedness between managing change, stretching ourselves, and stimulating innovation.

I am thankful for the excitement and collaboration provided to me by many industry professionals and friends from multiple companies in the Research Triangle Park in North Carolina. Their professional skills and willingness to step out of their comfort zones helped me fully understand the importance of training our educators to learn and pass on these skills to students.

Lisbeth Borbye
January, 2010

CHAPTER 1

The Comfort Zone and "Being out of It"

1.1 COMFORT

Being comfortable sounds like such a nice state. For example, we can be comfortable in our clothes, cars, houses, jobs, countries, or with other people. It provides rest, probably a sense of self-confidence and is not threatening at all, but safe. Feeling comfortable is something all people need to do regularly in order to obtain or maintain a balanced mind. But feeling comfortable ALL THE TIME may numb us and make it much harder to appreciate the feeling. It may also create complacency, boredom, and an ungrateful or negative attitude. One could draw parallels to hunger and life. Once hungry, being fed is a wonderful experience and the food tastes so good, but if overfed or never hungry, food essentially looses its value. If we lived forever, then our living hours would be so abundant that the value of life itself could be disputed.

> Being comfortable all the time
> may numb us.

1.2 BEYOND COMFORT

Getting away from our comfort zone is usually something we resist. Exceptions are people who are driven by extremely adventurous and risk-seeking personalities. Most of us thrive in environments and situations that seem controllable (by us), and we can most often be characterized as being change-adverse. There are, however, both voluntarily chosen and unavoidable changes we need to handle during our lifetime. For those changes, which can be considered of the category of actions leading to voluntary changes, there is a set of socially accepted responses and actions. Examples of such changes are marriage, birth of children, a new job, and taking out a loan from the bank. Although anticipated, these changes usually create some level of crisis; because we have seen others in similar situations, we can cope by following an accepted model behavior: adjusting to an adult spouse, taking care of infants, adapting to a new job, and paying debts can in many cases be done relatively easily.

It is much harder to deal with changes we do not choose for ourselves. Some of these are changes we have to anticipate and some are unexpected. Examples of anticipated changes, of which we have no control, are the timely loss of a parent, children moving away from home, and aging. Basically, these are all situations we can expect and we can prepare for them to a certain extent. Coping with these changes is not always easy but the unwritten laws of "normality" make these part of life's tapestry. Although we would rather be in control, we know these changes happen and must happen. On the contrary, unanticipated severe changes can be much harder to cope with. Examples are sudden accidents, disease, untimely deaths, loss of possessions, and scarcity of water, food, and safety, or maybe even winning the lottery. Most unanticipated changes, however, are much more mundane day-to-day surprises or obstacles, such as getting a present, receiving a promotion, having a flat tire, engaging in a simple dispute between friends, undergoing a change in a procedure, or experiencing loss of electronic data. There seems to be an abundance of opportunities to experience unanticipated changes.

Challenge and adaptation
help build character.

Most agree that there is a lot to learn from change, and that the experience has the potential to enrich a person's point of view about what is important and what is NOT so important. Challenge and adaptation are part of what builds character and make us appreciate the "good times" more. Learning to manage change and letting go of what was is a part of the life cycle. It seems to be an advantage to avoid judging change and to consciously practice expecting the unexpected. My claim is that if we are able to learn this for ourselves and teach our students how to do so too, then our capacity for developing better solutions and sharper minds is enhanced. This claim is based partly on personal experience and partly on the feedback I have received from my students.

CHAPTER 2

Exiting the Comfort Zone: Reasons and Impact

2.1 AUTHENTICITY

We typically follow a set of behaviors modeled by a group of people to fit into an environment, often termed "going with the flow." What this means is that we stay in our comfort zone and that we do not have any incentives to be different or noticed for anything particular. Remaining this way promotes continuation of an existing situation. Most people dream about more and may admire certain people who have displayed a way of doing and being more. Such role models typically stand out because they directly impact others by their performance (morally, visually, economically, or by other means), and by leading society in a new direction. The role models are authentic in regards to one or more traits; authentic is here defined as the individual being different than the group.

> An authentic thought
> can create a substantial change.

All people enjoy a certain amount of authenticity based on their uniqueness. Why are some people so highly authentic while others are not? It may lie in a difference in personalities and experience. People who are less inclined to be authentic, may be easily satisfied and accept their current conditions or unwilling or unable to take risks and embrace change. More authentic people may thrive through curiosity, entrepreneurship, and a wish to serve great causes or simply through getting other's attention. These people are likely to develop authentic ideas to meet their needs and goals, and public demand and personal reward incentives are strong drivers. It should be noted, that not all highly authentic people become known to others and that authenticity carries risks of both failure and fame. An authentic thought, product or method (matters which are often patentable) can create a substantial change if it is integrated into society. Therefore, it is of great importance that students and their professors are equipped to befriend their own authenticity and innovative mindset, and also that they are knowledgeable about the impact they can make by using it.

2.2 ESSENTIAL PROFESSIONAL SKILLS LEARNING

Being outside the comfort zone is equal to being in an unfamiliar element and coping with the new element requires learning of new skills. Such skills help to expand the comfort zone, encompassing what earlier was beyond comfort. This is important because rarely do people thrive entirely outside their comfort zone. Said in another way, the greatest skill is to make the largest possible comfort zone, that is, to be able to adapt to a variety of unknown events, or be able to be comfortable with being momentarily uncomfortable.

> Being out of the comfort zone creates a demand
> to cope with the new elements and learn new skills.

There are a number of so-called essential professional skills (see Resources, reference [1]), which can assist in expansion of the comfort zone and developing good coping mechanisms, such as interpersonal skills (leadership, teamwork, communication skills, humility, cross-cultural understanding), discipline, flexibility, creativity, ambiguity, change and expectation management, and ethical behavior. Often courses are offered in a few of these topics, but gathered together, they provide the fastest route to "zone-expansion." How can we provide training in all these disciplines simultaneously? The truth is such training already exists in an informal format and most people are unaware of it. It is called "on-the-job learning" (a parallel in life to this learning is simply "life learning" or "life skills"). Most people will be exposed to it by the sheer acceptance of a position. Its often unconscious existence renders it vulnerable to the ability of the individual to evaluate its outcomes and fully comprehend its meaning. The implementation of good coping practices prior to the onset of unexpected reality has enormous and positive consequences and is desirable both on the job and in life itself.

2.3   CHARACTER BUILDING

Moving away from the comfort zone means we need to adapt to something new, face challenges, and cope. It is during such times we make new choices and introduce a set of consequences for making these, and we learn more about our own character as it develops. Most likely, we are able to make better choices when we are in possession of a basic set of essential professional skills to guide us.

2.4   EXPANSION OF PERFORMANCE AND CAPACITY

After traveling to a new country, we may return home with new impressions and better understanding of other cultures. The same is the case when traveling outside the comfort zone. Stretching our imagination and mind helps expand our horizons. It seems likely that such stimulation may enhance our overall capacity and performance.

2.5   SURVIVAL

Sometimes, we are pushed out of the comfort zone. It is simply necessary to live there for a while in order to survive. This is often the case when losses occur, and we need to seek help, new ventures, homes, jobs or social relations. Such times are usually associated with much stress and essential professional skills can be a lifeline. The additional impact of surviving such scenarios is a deep insight and a capacity to help others in similar situations.

> Experiencing something new
> may enhance our overall capacity and performance.

In light of the current economic crisis, the topic of being out of the comfort zone and managing change is timely. People who are losing their jobs and homes need to find new ways to earn a living and places to live. Giving our students and professors an opportunity to practice what it is like to have to start over and "press the re-set button" seems more urgent and important than ever. Giving all of us a chance to learn to work together and share with each other in times of need is an essential skill that is priceless.

CHAPTER 3

Getting Educators and Students out of the Comfort Zone

3.1 COMFORTABLE CURRICULA AND BEYOND

Traditional teaching in schools and universities typically relies on lectures and exams. Often the same content is taught year after year. There is a focus on one-way communication and data-heavy, highly focused content. This content-oriented teaching style ensures that students are exposed to the expected academic details in a defined amount of time. As the years go by, professors may feel the need to update the content of their lectures in order to provide appropriate teaching and also stay interested in the materials themselves. Rarely does the teaching style change once instituted. Students are most often learning through passive assimilation of content, which often is disrupted by distraction and causes boredom. This is true especially in the current high-technology era, in which students enjoy continuous multitasking and multi-communicating.

> Professors can inspire by bringing topics
> into context and add unexpected elements.

What can be done to incite inspiration for both professors and students? A visit outside the comfort zone for both parties may be the answer. Professors can consider addressing new topics and giving away some of the control over the class-time to the students. They may think about using different teaching styles. Just like stand-up comedy, professors may elect to perform "stand-up teaching" and use a captivating storytelling style. They may choose to infuse their personality and humor into teaching. In addition, "reactive teaching" or an "open mike" style, in which professors respond immediately to comments or happenings in the classroom, can create an intense and interesting environment in which students and professors partner in learning. At first this may be awkward for both professors and students and require courage, self-confidence, and enthusiasm.

Giving students a way to have their voices heard and their contributions discussed is a foundation for sustaining their interest and learning. Professors can also ask more from their students by requiring varied and unusual assignments and assignment performance styles, such as teamwork. A particularly effective way to stimulate student interest and alertness is to bring topics into a real-life applied context, add unexpected elements to both teaching and student assignments, and to require discipline, teamwork, flexibility, ambiguity management, and innovation. This topic is described in detail in Chapter 4.

3.2   THE UNIVERSITY COMFORT ZONE AND BEYOND

Universities are worlds in themselves. There is a certain culture and language, and a specific set of rules for both students and professors. There is, of course, also a world outside the universities, for example, the industry world. Another culture and different rules exist here. Learning about a new environment is like learning another language. There is great potential in asking our educators to become partners in education with professionals from other environments, such as industry, government, and non-profit organizations. Many professors resist as it is truly away from their comfort zone, but once the alliance is initiated, they usually find it refreshing and stimulating. Getting to know the worlds outside the university may also provide a seed for additional collaboration and sharing of resources.

> Students who master essential professional skills,
> are likely to adapt quicker to change than other students.

Exposing students to industry or other non-university environments is an excellent chance to provide "on-the-job" education prior to the actual job situation. Students who have no job experience are often stunned by the differences between the university and the environments outside the university. Typically, they gain valuable insight in essential professional behavior. Examples are meeting on time, dressing professionally, staying calm and present and maintaining a positive "can-do" attitude. Projects in these environments are typically interdisciplinary in nature and are performed in a team setting. They require and foster a high level of interaction among employees with different expertise.

3.3   EDUCATIONAL COLLABORATIONS BETWEEN UNIVERSITIES AND WORK ENVIRONMENTS OUTSIDE THE UNIVERSITIES: INCENTIVES AND IMPORTANCE

External drivers exist for students and professors to exit their comfort zone. In the case of interacting with future employers students benefit by adding workforce-relevant training to their education and evaluating prospective jobs during their studies. Professors who work with employers can gain access to new know-how and resources. Employers can also reap benefits from collaborations with universities by learning how the university environment works and how to access the academic resources. Students and professors often envision the environment outside the universities as a "black box." Employers often have the same picture of the universities. By collaborating, it is possible to build synergy between the two "worlds." The most common incentives are shown in the table below:

Table 3.1: Common incentives for teaching and learning out of the comfort zone.

Stakeholder	Out of the comfort zone scenario	Common incentives
Students	OOC learning with employers and organizations as described herein and in Resources (references [1] and [2])	• Competitive and employment-relevant education • Essential professional skills • Making a difference during studies • Employer network • Employment readiness
Professors and universities	Educational collaboration with employers as described in Resources (reference [2])	• Access to employer resources and know-how • Potential for research collaboration • New type of network • Course and program popularity • Fulfillment of university mission
Employers	Educational collaboration with universities, professors and students as described in Resources (reference [2])	• University alliances and access to university resources • Fresh (and free) out-of-the-box work • Training of students (future employees) in matters of importance for company including flexibility, discipline, and teamwork • Evaluation of students for future employment • Access to potential employees • Service to society

3.3.1 STUDENTS

Students want competitive and employment-relevant education. Much of the success on the job depends on mastering essential professional skills, and, therefore, it is important to make learning of these skills available to students. Many students aspire to make a difference during their studies and feel they can do so by creating value for an employer while working on a project with a focus larger than themselves. It provides a feeling of co-existence and interdependence rather than a lonely pursuit of individual resource building. One very large benefit is the employer network and employment readiness students can acquire through their interaction with potential future employers. The largest obstacle to students is the lack of appropriate degree program offerings.

3.3.2   PROFESSORS

One highly treasured asset is the access to employer resources and know-how professors can gain via interaction with the outside world. In addition, the interaction in itself provides a platform for new research collaborations. Expanding the interaction with employers outside the university is often part of a university's mission. Professors also gain from providing exciting new courses and programs leading to higher enrollment and popularity among students. The barriers to success are multiple and typically based on a reluctance to change and a lack of understanding of the value the interactions can contribute. The fear of doing something new and knowing how to contact non-university working professionals is a common challenge for professors. Fortunately, at most universities, some professors are already familiar with industry collaborations and can assist inexperienced faculty in their endeavors.

3.3.3   EMPLOYERS

Gaining university alliances and access to university resources is of interest to most employers. The often free student labor and out-of-the-box thinking can be a substantial value-addition. Employers also benefit from exposing students to their work environment and training them in matters relevant to future employment. Both parties have opportunities to evaluate each other. Many employers also wish to give services to society and training students is one way of doing so. Employers need to know that universities are seeking to develop educational alliances with them and that intellectual property matters are resolvable. Another challenge is an often highly dynamic resource and logistics situation.

Flexibility and adaptation are needed when change happens, and it does so often in environments outside the university. Students who have been introduced to these environments usually display an improved attitude and interdisciplinary understanding. It is evident that graduates, who have learned essential professional skills, adapt quicker to the highly dynamic professional environment than those who have stayed in their comfort zones. Chapter 4 outlines various ways to train students to cope with typical changes after simulating these.

CHAPTER 4

Principles of "Out-of-the-Comfort-Zone" (OOC) Teaching

4.1 EXISTING AND NEW CURRICULA

4.1.1 CONTEXT CREATION

Putting learning in its context is one of the ways education can "come alive." Students tend to be interested in why they are learning certain things and how the learning can be applied. This is true for both existing traditional and new curricula. For example, knowing what the end product of a certain process is and what impact the product has are as important as knowing how the process to produce the product works. Embedding a topic into real-life scenarios to which students can relate, helps create understanding for why it is important to learn about these. With a little effort, existing curricula can be "transposed" into a given context. Alternatively, this goal can be reached by the creation of new curricula and the use of immersion learning (see Resources, reference [2]). This type of learning requires that students are immersed in an environment in which the topic of study has great importance and interest, and in which, they automatically gain exposure to the context. Student internships or industry projects can be considered immersion learning vehicles if they are conducted in the industry environment. Immersion projects most often require interdisciplinary resolution. When students are surrounded by a culture in which these requirements are the norm they are encouraged to develop integrative problem-solving and innovation skills and to take responsibility for the outcome of a project.

4.1.2 CHANGE INTRODUCTION

The most prominent element of OOC teaching and learning is change introduction and management. What is the advantage of introducing change and the idea behind simulation of real-life situations?

> Crises provide valuable
> pedagogical opportunities.

During our change management process, we strive for optimal scenarios that include the new parameters, rather than giving up because something changed. Such important skills can be practiced in school. Instead of letting students complain about changes, professors can elect to require good change management skills and flexibility from their students.

The basic teaching method relies on creating certain manageable crises for the student. This brings them out of their comfort zone and provides ambiguity, distress, and sometimes a great deal of discontent. Initially, students may be confused, uncomfortable, and feel unfairly treated (which they are), but it can very soon be followed by curiosity and a feeling of "freedom to operate" and will to pursue excellence despite the odds. The goal is to make students understand that no matter what happens, they always have to "put their best foot forward." The only alternative is to quit, an option that is not realistic if a student wants to fulfill the requirements to pass a course. It is, of course, always an option on the job, but with the above-mentioned conditioning, graduates will be more likely to understand what happens and know when and when not to take things personally.

The following four sections describe ways to mimic real-life on-the-job common scenarios of introducing change for teams, projects, instructors, and performance indicators. These are examples which can easily be applied to any project-based curriculum. They are by no means exhausting the overwhelming amount of change management opportunities that exist in the industry and other immersion environments. Each table shows the unit that is subjected to change, the actual change and the situation it may mimic, and additional suggestions to change (and distress) inducers; it is followed by a description of the individual components. Students have an opportunity to practice all of the essential professional skills mentioned in Chapter 2 using the indicated methods for change.

4.1.2.1 Teams

Terminate Existing Teams   It may seem very radical and "unfair" to terminate certain teams and not others. It promises to make students upset until they understand the meaning behind the act. Terminating a team mimics a number of common situations such as a re-organization, change of strategy, a lack of value of the team product, or simply lay-offs. Teams should have time to work together before being terminated and there must be a plan to place terminated teams and team members in other continuing teams. This strategy has the additional effect of introducing an "advantage" for students who are on a continuing team, whereas the displaced team members must work harder to acquire a similar level of skill if team projects are different from each other.

Start New Teams   Displaced teams and team members can be asked to form entirely new teams, thereby starting over. Students may initially feel disadvantaged compared to their peers in continuing teams. Because this situation may mimic a new product interest and maybe a change in strategy, it may soon be perceived as a very positive and selective event because a company often seeks to pick the very best people to develop an entirely new idea.

Shuffle Team Members to Other Teams   Moving individual team members promises to create both confusion and discontent for the same reasons described above. Students get an extraordinary oppor-

Table 4.1: Changes for teams.

Unit	Change description	Examples of real-life situation simulation	Additional change inducers
Teams and team members	1. Terminate existing teams 2. Start new teams 3. Shuffle team members to other teams 4. Merge teams	1. Re-organization of company 2. Product termination 3. Change in strategy 4. Lay-off 5. New product interest 6. Merger	1. Rank team members 2. "Fire" team members
Team member roles and responsibilities	1. Alternate team leaders and team members 2. Ask more of some students, less of others	1. Performance and skill-based promotion or demotion 2. Task-based responsibility assignment	1. Give weaker person team leader role 2. Combine personalities which do not match 3. Create additional opportunities for conflict and conflict resolution

tunity to display flexibility and a positive attitude during such transitions. The purpose of shuffling is to instill a "can do" and "can do anything you ask of me" attitude. Re-organization and strategy change as well as "cherry-picking" the best talent are some of the situations this mimics.

> Sometimes a seemingly bad situation
> turns out to be a good one.

Merge Teams Merging two or more teams into one simulates what often happens when companies merge or acquire parts of each other. People from similar teams are asked to define one set of rules and erase differences in order to create one set of rules, deliverables and vision. Mergers may seem counterintuitive to growth because growth often means expansion and initiation of more teams. It is an exercise with demands of adaptability and flexibility.

Alternate Team Leaders and Team Members Teams usually have a team leader and team members, who report to the team leader. Such hierarchy is normally somewhat stationary, but changes may occur based on performance and skills needs. Changing a student's "position" from being the team leader to being a subordinate will surely be a challenge for the entire team and good practice in respecting each other regardless of status. It is important that all students try both roles either in different projects or by using this tool.

> Learning to walk in each others shoes
> may foster better teamwork.

Ask More of Some Students, Less of Others   Some people have great skills in an area where others do not. Therefore, it is natural to ask the more skilled people to perform a specific task mimicking such skill-based responsibility assignment. Sometimes this can create a skewed distribution of workload. It may seem like a person is carrying too heavy a load and perhaps so. This person may continue to do so, or start to educate team members in this area and delegate some of the responsibility. It may even be necessary for this person to compromise on quality and learn about resources, quantity, quality, and time, which are the interconnected parameters of project management. Students benefit greatly from being tasked to resolve such differences.

Rank Team Members   Ranking performance may create internal competition and be stressful for some people, but for others, it is an incentive to perform at their very best. Employees typically get ranked. The rank an employee receives translates directly to compensation and promotion issues. Being ranked is familiar to students through the grade system, except no two people can share the same rank.

> Each team member has to add positive value
> to the team spirit and the project.

"Fire" Team Members   Dismissing a team member can be necessary if there is an attitude or discipline problem. Such problems can be prominent and examples are contagious negativity or an inability to show up on time, produce the necessary deliverables, or act appropriately to other requested actions. Typically, a series of warnings are given before such action is taken. Once "fired" a student should not receive a passing grade for the project. Alternatively, dismissing a student from a team can be done with the only purpose of mimicking a lay-off scenario in a company. In this case, the dismissed student should receive a passing grade. The important learning relates to being able to survive a lay-off and not to take it personally (unless there is a reason to do so, as indicated in the above-mentioned scenario).

Give a Weaker Person Team Leader Role   Students who are natural leaders tend to want to lead teams. Such students can learn a great deal of humility by experiencing the role as a subordinate, especially a subordinate of a less strong person. Likewise, a student who would rather be a subordinate than a leader can learn a lot from having to lead others. In both cases, the parties are getting out of their comfort zones.

Combine Personalities that do not Match Assembling teams with different personalities is sure to produce interesting work at the interpersonal level. Combine the "big picture" person and the detail oriented person, the extravert and the introvert, the leader and the follower. This situation mimics real-life at any place. Students have to learn that they cannot pick their colleagues (unless they are in charge of their own company), and they have to learn to work well with everybody.

Create Additional Opportunities for Conflict and Conflict Resolution There are many other ways change can be introduced. Permutations of the above-mentioned elements can be used. In addition, teams can be forced to adapt to new management and cultures, etc. Imagination is the only limit.

4.1.2.2 Projects

Terminate Project The termination of a project mimics re-organization, a strategy change, lay-offs, or perhaps ambiguity regarding the strategy. It is possible that the project has not provided enough data at a given deadline to assess the viability of the project (hence ambiguity). This is not uncommon, and good companies will devise a way to either attain the necessary data in a timely manner or eliminate the project. Students tend to understand these alternatives clearly, once explained.

Start Alternate Project Teams may be asked to start over in a new focus area. The effect is usually an immediate reaction of "unfairness" and difficulty letting go of potential results from the old focus area. The reasons to start alternate projects are often the same as those involved in terminating projects but with the intention of keeping personnel employed.

Make Projects Compete Competition between teams or between companies may also be a reason to start new projects, and it may be enhanced by starting multiple teams with identical projects. This process is used to increase the speed of the work done and potentially identify the best employees within a given focus area. Students relate well to this procedure because it is similar to the competition for grades.

> Competition, idealism, and unmet needs
> are drivers for excellence and innovation.

Require Innovation One way students can be "stretching themselves" intellectually is through innovation. Usually, students find it very exciting to be allowed to create something new with a positive impact on society. It is an essential contribution in companies and critical for the development of new products and their conjunctive intellectual property. It is helpful to make inventions part of the deliverables. To get top grades (mimicking a top ranking or a bonus) students could be required to attempt to invent a new product.

Table 4.2: Project changes.

Unit	Change description	Examples of real-life situation mimicking	Additional change inducers
Project	1. Terminate project 2. Start alternate project 3. Make projects compete 4. Require innovation	1. Re-organization of company 2. Performance-based layoff 3. Strategy change 4. Strategy ambiguity 5. Competitiveness issue 6. Need for intellectual property 7. Need for new products	1. Provide much ambiguity 2. Provide little guidance 3. Make projects open-ended
Project deliverable	1. Change content 2. Change format 3. Change delivery method 4. Change delivery person	1. Dynamic communication 2. Strategy under development 3. Management team unsure 4. Change in audience 5. Change of other logistics 6. Delivery person not available 7. Delivery person deemed incapable	1. Prepare to present projects in front of industry professionals 2. Limit resources 3. Require adjusted project plan
Project logistics	1. Change location for project work 2. Change time-line for project 3. Change dead-line for project deliverables	1. Sudden change in resources and availability 2. Term change 3. Rush order	1. Require adjusted project plan

Change Project Content   After teams have provided the first report in a focus area, the content of the project may change slightly, but not completely. This may foster some frustration and questioning of the leadership, especially if it occurs repeatedly. This situation mimics that of a dynamic development and decision process. Often leaders do not know exactly what they want at the onset of the project, and it is through a partnership with the development team that the vision and strategy are refined. It is also possible that the audience for the project may change and that the content level needs to be adjusted accordingly.

> A need for adjustment is an opportunity
> to create the optimal outcome.

Change Project Format   There can be a reason to change the format, for example, having a project in the early development stages. If the project is received positively, or if other projects fail, the scope may expand. Outside drivers may also play a role, such as new strategies in intellectual property management, or simply market forces. Also, the project may demand both teamwork and individual deliverables. The final product may be one or more written, oral, or electronic presentations. Students learn to display flexibility by being asked to change the format of their projects.

Change Project Delivery Method   If there is a change in the size of the audience or time available for delivery, it may be necessary to adjust accordingly. For the same reasons of a dynamic development process as mentioned above, multiple other aspects of the project can change: One example is the project delivery method. Instead of an oral presentation, a written business plan may be chosen or both may be desired. Requesting such changes very close to the time of delivery mimics realistic last minute decision-making in companies.

Change Project Delivery Person   In the event of an unexpected event such as illness, accidents, or traffic delays, the "show must go on" and a "back up person" must deliver the project. Also, there may be other reasons a person is deemed incapable of presenting a project deliverable. Students are often surprised about the fact that an event often does not get cancelled or postponed, but, instead, relies on another representative from the team. Because time is an extremely valuable asset in companies and people need time (and get compensated) to arrange meetings, the economic loss for canceling a meeting is substantial. Such learning helps the students ensure that all team members understand the project and are capable of presenting it.

> Thinking about a "plan B," especially
> in situations of high visibility is a good idea.

Change Location for Project Work   Another method to introduce a way to practice flexibility is by requiring students to show up at a different location. It can be in a classroom, a company site, or maybe a field visit. Getting used to functioning in different locations outside the regular comfort zone (which often is a bench, a personal office, or a cube) helps students adapt quickly to both anticipated and unanticipated events, such as guest visitors claiming office space, inspections, re-locations, and building closures.

Change Time-Line for Project   Discipline is necessary to conduct projects on time and with the necessary quality. Letting students understand how to juggle deliverables in terms of their quality, quantity, length of production, and the resources needed for production introduces them to real-life project management. When time-lines change, other things also have to change. Students normally become dismayed if the timeline for a project is shortened rather than lengthened. At the same time, they learn to optimize their delivery process.

Change Dead-Line for Individual Project Deliverables   Some students' deliverables may get a shorter dead-line than others. Later, these and other deadlines may be changed repeatedly. There are a variety of ways this can play out, and everything mimics the dynamic environment in a company where some deliverables are needed suddenly while others are not needed.

> An earlier than expected deadline
> means a faster route to the end must be designed.

Provide Ambiguity   When students know exactly what to do, the process they follow is part of their comfort zone. Removing some the process parameters and leaving students to find out how to attain the deliverables gives them an opportunity to enhance their skills. In addition to their usual result-orientation, they need to find out exactly what they are supposed to find out. This task demands pro-active interaction with all potential stakeholders and consultation of a variety of media. Students need to inquire about whether or not the context and process are built correctly to address the issue and provide multiple recommendations for continued amendment and resolution.

Provide Little Guidance   Chief Executives in a company often have little expert knowledge about particular product details. Their job is to organize the vision and "large picture." Employees who are experts in the focus area may spend time educating their leaders about the product and can expect little or no guidance from their superiors. Students are unfamiliar working with little guidance. Many of them initially rely on consulting basic internet resources, but soon learn that much more depth is needed. Learning how to obtain in depth information from experts and media and assemble and analyze these help students become independent.

> Ambiguity leaves space for risk-taking,
> independence, and individual decision-making.

Make Projects "Open-Ended"   When projects do not have a clear end-product, or if they have more than one possible solution to a problem, students have to satisfy both scenarios. They must envision solutions and be able to prioritize. Such projects demand mastery of both context and context details and can help students appreciate how much effort goes into developing entirely new products and solutions.

Prepare to Present Projects in Front of Industry Professionals   Presentations prepared for a university audience are usually saturated with data and details. Depending on the audience in a company setting, students will have to evaluate and adjust their presentation style and content carefully. Generally, such presentations need to visualize the "bottom line" and not have too much detail. It is desirable to mimic the company situation by asking students to dress and conduct themselves professionally, and particularly helpful if students can present on the company site.

Limit Resources   Yet another way to make projects challenging is to limit the original resources available. This mimics situations in which a company finds itself in a sudden crisis. When students have less access to resources (budgets, equipment, time, knowledge, partners, etc) it forces them to be both creative and to make hard choices.

> When conditions are less than
> favorable, creativity grows.

Require Adjusted Project Plan   Student may be asked to submit an adjusted project plan whenever a change has occurred. Multiple changes may occur during a project life cycle. This takes time away from the project and often causes some frustration. It mimics the scenario of a company seeking to document all changes as a part of the job.

4.1.2.3 Instructors
Change Instructors   Re-organization occurs frequently in companies and very seldom at universities. Often, it leads to the formation of new teams and also new leadership. Students are used to "reporting" to the same instructor during a project. Different instructors can lead different parts of a project or instructors can shuffle at random times during the project. By changing instructors during a project, students learn to adjust to different management and leadership styles.

Make Instructors Unavailable   Not knowing what to do or who to ask for information is incentive to seek advice from alternative sources and create a plan for what to do. Students gain valuable experience and independence by exposure to intermittent lack of leadership, while they depend on getting their work done (to get their grades). A period of initial helplessness can be expected. Making instructors less available shows a common scenario when leaders have sudden travel, audits, or other responsibilities, which hinder interaction with employees.

Table 4.3: Instructor changes.			
Unit	Change description	Examples of real-life situation mimicking	Additional change inducers
Instructors	1. Change instructors 2. Make instructors unavailable 3. Make instructors hypercritical 4. Make instructors hypocritical	1. Re-organization 2. Lay-offs 3. Sudden unavailability 4. Micro-management 5. Macro-management	1. Provide ambiguity 2. Provide little guidance

Make Instructors Hypercritical   Some leaders are very detail-oriented and want everything exactly the way they envision. They see one solution per problem. Teams may experience the art of finding a balance between the usual ambiguity that surrounds a project and provision of a product of the leader's exact vision (and such a vision can be dynamic). It can take a long time to get it right and intermediary versions will most likely endure much criticism. This is not unlike the academic environment, in which students and professors are conditioned to reconstruct and critique to pursue the absolute true answers.

Make Instructors Hypocritical   Other leaders are most concerned about the big picture and less interested in the details. They anticipate that each problem can have multiple solutions and attempt to nurture the best possible solution. This may be an even harder scenario for some students because it demands a higher level of independence and sense of quality control. It may also be more difficult to assess when a successful product is made and how to measure the success. Many top leaders are interested in things working well but not necessarily how they work. For students, it is extremely valuable to try to work under such conditions.

> Some leaders worry about the details while others do not, and it is important to recognize both work styles.

Provide Ambiguity   Working with hypocritical leaders is an assurance for working with a high level of ambiguity. This scenario can be mimicked by requesting students to produce several weighed recommendations between which the instructor eventually will choose.

Provide Little Guidance   Students may work with very little guidance from the instructor to accentuate the situation mentioned above. They may have to seek help from alternative sources.

4.1.2.4   Performance Evaluation

Table 4.4: Change in performance indicators.

Unit	Change description	Examples of real-life situation mimicking	Additional change inducers
Performance evaluation	1. Evaluate attitude and professional skills, academic learning, and technological skills 2. Require self evaluation 3. Require 360 evaluations	Typical company evaluation containing self evaluation, evaluation by supervisor and 360 evaluation	1. Make teams evaluate team members together verbally 2. Ask for commitment to future change and ways to measure successful change
Student grading	1. Let instructors grade using company instruments 2. Let professors finalize grade		1. Let students participate in grading each other

Evaluate Attitude and Professional Skills, Academic Learning, and Technological Skills   Students are used to being evaluated on their academic knowledge; interpersonal skills and other professional skills do not matter as long as their assignments are delivered on time. To mimic a performance evaluation in a company, students' academic knowledge, technological skills and professional skills must be evaluated concomitantly, thereby providing incentive for students to master and demonstrate these skills. It usually takes a while for students and professors to learn that professional skills are as important to success as academic knowledge and technological skills.

Require Self Evaluation   Knowing oneself is imperative to success. A self-evaluation gives a picture of how a person views her or himself. It does not necessarily reflect how other people view the person. When students have to evaluate their own performance and address whether or not they have met the goals they set forth at an earlier time, they may develop a structured performance style and improve their goal setting.

> Getting a realistic picture of
> oneself is helpful.

Require 360 Evaluations   Evaluating a person in a 360 perspective means that all people who interact with the person review this person's performance (including the person him or herself). This is a very powerful tool and also indicates that employers are evaluated by their employees. Therefore, to transpose this to students, instructors, and professors means that students evaluate instructors and professors and vice-versa, something that may be out of the comfort zone. This method provides a way to find pervasive patterns of behavior and performance as well as rare and extreme sentiments.

Let Instructors Grade Students using Company Instruments   Instructors may choose to use standard company evaluation methods and instruments. Some of the performance indicators are very different than those used at universities, and examples include elements such as teamwork, professionalism, being on-time, being well-prepared, attitude, willingness to assist others, etc. Each element represents a percentage of the total evaluation. Students are very adept at improving their grades and will soon seek to excel within these measures of success if these are part of the grading regimen.

> Make essential professional skills acquisition
> part of the evaluation and students will excel in these.

Let Professors Finalize Grade   Some uncertainty about the final grade can be added by using both academic and company evaluation methods. In this way, students will seek to stretch themselves in all areas. For reasons of transparency, it is important to let students know beforehand how the evaluation of their performance will be conducted, what the assessment parameters are, and how they are weighed.

Make Teams Evaluate Team Members Together Verbally   Verbal "confrontation time" (time spent together while confronting each other with certain issues) is a very effective way to provide feedback when using good communication and conflict resolution skills. It requires that the person overseeing and guiding the evaluation stays emotionally detached. This type of evaluation may at first seem uncomfortable for students, but when conducted properly can lead to successful interpersonal and personal goal setting and relief for the entire team if conflicts have been hampering teamwork.

Ask for Commitment to Future Change and Ways to Measure Successful Change   A new and sometimes uncomfortable request is that of future commitment. Students often think very linearly and when a project is finished, it may be forgotten quickly. However, future commitments that may play a role in a new project last, and may be a result of performance and accountability in an earlier project. This is the case in most company settings where performance is evaluated regularly and growth is measured by review of anticipated growth (which was committed to at the previous performance assessment) and actual growth.

> Future commitment is a driver
> for personal growth.

Let Students Participate in Grading Each Other   When students grade each other, professors often experience them as their own harshest judges. Done orally and in a team, one can be sure that nothing is left unnoticed. Students may want the highest level of excellence and in voicing criticism indirectly expressing how they would have performed given the chance to do so. A pitfall is the scenario of rivalry and competition, in which students want other students to be evaluated more negatively than themselves. It is important that professors balance students' criticism of each other with positive remarks. Another important aspect to be aware of is that of "performance inflation" where no student wants to say anything negative for fear they will be judged harshly in retribution. One way to counter such sentiments is to change the tone of the language and make students express strengths and opportunities for improvements rather than positive and negative remarks. In addition, being able to present strategies for optimizing and improvement of each person's performance can be made part of the project.

4.1.3   RESPONSE TO CHANGES

Students who are unfamiliar with change management may respond with confusion and distress when confronted with high levels of ambiguity and unexpected changes in team compositions, projects, instruction resources, and performance indicators. Their initial view of the instruction process is typically narrow and situations of change within the process met with irritation. As students progress in their understanding of the need for change management and the value it brings to project and life contexts, they most often embrace an increased level of flexibility. Students who have been repeatedly exposed to the teaching methods described in the section "Change Introduction" grow from displaying the necessary flexibility to expressing excitement. They adjust to new parameters and develop their own understanding of how to solve a problem or design a project process. They courageously expand their comfort zone by taking steps out of it. Professors are privileged when they have the opportunity to follow this development and can practice taking similar steps themselves alongside their students.

4.1.4   MONITORING THE COPING PROCESS

During the development from distress to understanding students must be monitored carefully. Some students have a very small comfort zone and will need much practice to expand it, while others are more prone to want to try something new. A change in student's body signals can show a great deal about how comfortable he or she is. Professors need to assess if such signs are caused by the new situation or are permanent. Examples of signs of distress are a high-pitched voice, sweat, tears, nervous movements, bad posture, and poor eye contact. Other signs are confusion, frustration, resentment, and a hopeless attitude. Coherence in communication and performance and the energy that the student displays are other indicators of the comfort level.

Professors can create a "safe" and inclusive atmosphere by providing anecdotes describing how former students have been in similar situations of distress. In this way, students can think of distress as a normal reaction to a situation that is new to them. If students experience more than average distress it is very important that professors are prepared to guide them to a less distressed state of mind.

> Praise and celebration of improvements
> stabilizes self-esteem.

4.1.5   MENTORING AND STUDENT DEVELOPMENT PLANNING

Former students are usually in the lookout for the particular learning, which will result from crisis management. They know there is an educational reason for the crisis, and they are prepared for the unexpected with the understanding and trust that there is a valuable lesson to learn. Therefore, former students can provide excellent mentors for students who are new to the teaching method during the introduction of manageable crises.

I recommend that professors take time to assess and mentor students at least once per semester and help them define their strengths and weaknesses as well as a path to improvement and ask them for their commitment to walk this path. In this way, students can plan their actions and monitor their improvement throughout the semester. Students may make this formal by signing their plans to simulate how self assessment is often done in companies. A plan could contain the following items: student name, semester, mentor name, project name, project location, project subgroup, project deliverables, recent successes, current challenges, student strengths, student weaknesses, plan for improvement, commitment to improve, and student and mentor signatures and date. An example of a student plan is given in Table 4.5.

> Making an effort to praise students' achievements
> is as important as correcting and critiquing their work.

4.1.6   REFLECTION ON LEARNING

Creating awareness of what happens followed by reflection of the consequences thereof is a way to speed up the process of coping and to enhance learning. One could say that what happens may constitute an unconscious way of learning, while reflection about what happened provides insight at

Table 4.5: Student plan of commitment. *(Continues).*

Student name	Matthew XX
Semester	Fall 20XX
Mentor name	Daniel XX
Project name	Greening the Workspace
Project location	Company XX
Project subgroup	Laboratory Spaces
Project deliverables	Plan for reduced use of non-recyclable and recyclable items
	Plan for recycling
	Plan for new use of non-recyclable and recyclable materials
	Plan for minimized carbon and electricity expenditure
Current successes	Teamwork going really well, everything is on schedule and exciting. We are able to come up with new ideas for the company
Current challenges	Most recent presentation did not go well. The presenter (I) was too nervous

Table 4.5: *(Continued).* Student plan of commitment.

Student strengths	Very thorough and consistent work and entrepreneurial, great interpersonal interactions and teamwork
Student weaknesses	High-pitched voice when presenting, very nervous
Plan for improvement	1. Practice presenting in front of mirror at home
	2. Try to speak with different pitches to find one that sounds confident
	3. Practice entire presentation in front of team members
	4. Also practice having good posture
	5. Eat well and exercise the day before a presentation
	6. Get enough sleep
	7. Think about what is the worst that can happen and that this will pass too
Schedule for plan	I will practice at least once per week during the entire semester
Commitment to improve	I commit to follow my plan of improvement
Student signature and date	Mathew XX XX/XX/20XX
Mentor signature and date	Daniel XX XX/XX/20XX

the conscious level and thereby an opportunity to learn twice, optimizing learning by using different methods. Consciously knowing what is learned and what to do with this learning is a powerful tool for expansion of the comfort zone.

> Reflecting about learning is a way to
> accentuate the knowledge gained.

Reflection about learning can be part of the requirements to pass a course. It can be applied to various areas such as traditional academic knowledge, professional skills needed outside of academia (such as essential professional skills described in this work and technological skills related to a specific industry, if applicable), and at the personal level. Professors and instructors may want to point out to the students when they are learning specific skills because it may not always be obvious to the them.

Table 4.6 shows learning in the three different categories. In each case the following questions are posed: What did you learn? How did you learn it? Why is this learning important? How will you use this new learning? Reflecting about learning helps students define their strengths and weaknesses and understand better what needs to be addressed in their student development plans described earlier. Examples of student learning and reflection about learning are presented in Chapter 5.

Table 4.6: Reflection on learning.

Level of learning	Question and student answer for each level of learning
1. Academic skills	1. What did you learn? I learned that …….
2. Professional skills	2. How did you learn it? I learned this through …….
3. Personal learning	3. Why is this learning important? This learning is important because …….
	4. How will you use this learning? I will use this learning to ……..

4.2   LEARNING TO EXPECT THE UNEXPECTED

4.2.1   TEACHING MODE

Both students and professors can draw advantages from continuous change in teaching modes. Not knowing what to expect can stimulate student alertness and flexibility, while professors use their creativity to teach in multiple ways. Some examples of variety include requiring both teamwork and individual work, teaching through lectures, guest seminars, discussions, question and answer sessions, interviews, online interaction, introducing "real-life" context through visits to companies, field trips, making students "teach" through presentations, and practicing essential professional skills acquisition by introducing manageable crises as described earlier. Students will increase their alertness if modes are changed with short notice.

> When a solution to a problem is non-obvious or dynamic,
> students must first break the problem down to manageable parts,
> before they can solve the entire "puzzle."

A particularly interesting scenario is when a professor does not know what to expect. This situation can occur embedded in a well-planned class if the method of "reactive teaching" is applied. The professor reacts to student comments, questions, and immediate needs in a manner that may provide an integrative academic, professional and life skills context. This is different from the traditional question and answer session, which typically focuses on concise academic learning only, by the additional and prompt (unexpected) relation to context, importance, and impact of the issue addressed by the students.

4.2.2   ASSIGNMENT AND REWARD PARAMETERS

When assignment parameters are multiple and variable, students may end trying to find out a "system" of getting the highest grades. This is because such a "system" appears very dynamic, too dynamic from which to create sense. On the other hand, one may argue that there is only one "system" available to students and that the "code" is to do their very best at any time given the changing circumstances.

Part of the dynamic assignment structure may be the open-ended nature of the project subject or problem the students are trying to solve. If both professors and students are willing to let creativity and opportunities unfold along the way, there is increased probability for new innovation and unexpected and better products. In addition, both parties develop the necessary skills of prioritization and selection of likely solutions.

> Not knowing what to expect
> creates a focus on our ability to respond based
> on our prior experiences and current character.

Creativity and open-endedness may not seem as "safe" as solving a problem with a concise answer for most students. The incentive for students to embrace taking the risk needed to suggest solutions without fully knowing the consequences is a revamped performance and grading system, in which the skills described above are rewarded. Students also benefit from receiving public and personal recognition in celebration of their achievements. A prize or honor can work as an extra incentive. In contrast, it is necessary to include consequences for inflexible and other negative behaviors in the grading system.

> Professors need recognition
> for teaching in innovative ways.

Incentives for professors to embrace open-ended teaching styles may include review and reward of educational innovation including essential professional skills teaching. A review ensures that the new components are accepted and included as part of a professor's evaluation. Rewards can be given in various formats, such as a raise, an award, or other kinds of recognition.

4.2.3 PEER INTERACTION

When students work in teams, there are demands of their interpersonal skills. Conflicts are bound to arise and they often become more numerous if changes are introduced regularly, especially if these changes relate to the roles the team members have. Students become increasingly comfortable resolving new and unexpected issues during such challenges.

> Students benefit from exposure
> to conflict and crisis management.

With the appropriate performance indicators, students will seek to bring the best out of their team members and ensure that the team as a whole succeeds, no matter what the situation demands. They may also learn to embrace an assertive and mindful communication style in order to optimize understanding, avoid conflicts, and thereby increase their effectiveness by saving valuable time formerly immersed in conflicts. Teams may first need exposure to conflicts to reach such conclusions. Teams may also develop their own incentives for excellence and punitive systems for any non-productive behavior.

CHAPTER 5

Anecdotes of OOC Learning

The following anecdotes show how opportunities for essential professional skills learning can occur spontaneously. Each section is described through a student's perspective and ends with an example of reflection on learning. Please note that only the categories of professional skills learning and, the often overlapping, personal learning are included in these examples. Academic skills acquisition can be reflected upon in a similar manner but is not the focus of this work.

5.1   LIFE IS UNFAIR: CHANGE OF TEAMS AND PROJECTS

LEARNING: FLEXIBILITY, TEAMWORK, PRESENTATION SKILLS, AND POSITIVE ATTITUDE

The semester had just started and I really wanted to do well. It was my first time working with a company project and the instructor talked about the possibility of doing some teamwork. Initially, all of us were working individually on our own projects. I guess that made it easier to assess how good we were on our own. Also, I really did not mind because I was very good at what I did and I was not used to teamwork. Others had told me that sometimes people do not do their part, and I knew that would irritate me. Also, I worried that others may not be as good as me and that a team grade could drag me down to a lower grade than I deserved. All in all, I was pretty happy working on my own.

My happiness did not last long. I was shocked to learn that after 2 weeks we had to present our projects and then only half (8 out of 16) of these continued. This meant that the other 8 projects were terminated. Mine was terminated. After that I had to work together with one of my class mates on her project. It seemed unfair because she had a two week advantage compared to me. And it got worse. After yet another two weeks the scenario was repeated. Only 4 of the 8 projects were continued and 4 were terminated. Needless to say, I was angry when our project was terminated (again) and we were told to work together with one of the 4 continuing groups. Truly, it was unfair that the other 2 people had a lot longer time on the project than us. I did not hide my dissatisfaction when the instructor asked for feedback. After discussing my feedback with the instructor, I suddenly understood that there was a point in all of this. Reflecting on my learning helped me see that I had learned much more than I thought.

REFLECTION

Level of learning	Question	Student answers
Professional skills	What did you learn?	I learned that projects often get terminated in industry and that it is not always obvious for the people who work on these projects why it happens. I also learned that I have to comply with these decisions.
Professional skills	How did you learn it?	I learned this because my project was recently terminated.
Professional skills	Why is this learning important?	This learning is important because not all projects are viable and a company needs to "weed out" and prioritize among projects. There may be a number of reasons for that.
Professional skills	How will you use this learning?	I will make sure I understand my company's mission and if my project stays viable or not. I will be prepared to continue doing the (new) work that is assigned to me.
Personal learning	What did you learn?	I learned that I got very upset when my project got terminated after I had put a lot of effort in it.
Personal learning	How did you learn it?	I learned this because my project was recently terminated.
Personal learning	Why is this learning important?	This learning is important because projects get terminated often in industry and I have to show flexibility in order to keep my job.
Personal learning	How will you use this learning?	Next time my project gets terminated I will not get upset but be grateful that I have a job. I will adapt to new responsibilities quickly.

5.2   NO ROOM FOR ME: CHANGE OF PROJECT PRESENTATION LOGISTICS

LEARNING: TRUST, HUMILITY, POSITIVE ATTITUDE, AND LEADERSHIP

Something unfortunate happened when our team was supposed to present our projects to the company executives in their conference room. Our professor had no idea that so many of the employees wanted to listen to us. It meant that there simply was not enough room for all the people and then us! To be concise, it meant that some of us students had to be escorted to the mail room (far away from the conference room). Not great. Here we were, all of us "experts" in different aspects of the project. Basically, only half of us were in the conference room to "defend" our work. We were dressed up professionally, and we had been looking forward to being part of this. We in the mail room were also worried that our team members would not be able to answer all of the questions and do things as well as if we were there too (and yes, we heard later there was a question for one of the "experts," who was sitting in the mail room). It was a bit like being a second-range student. That aside, the presentation fortunately went really well, and the company executives were very impressed with us. They decided to perform more research on two of the products we presented. I think they saw everything as a team effort and did not pay too much notice to the fact that we were not all able to be there for the presentation.

REFLECTION

Level of learning	Question	Student answers
Professional skills	What did you learn?	I learned that an audience can be larger than expected, and regardless of this, one must comply with current fire codes.
Professional skills	How did you learn it?	I learned this when half of us were excluded from the conference room because many more than expected employees wanted to listen to us and half of us had to wait in the mail room.
Professional skills	Why is this learning important?	This learning is important because it is often hard to gauge the size of an audience. Also, another situation could be that an appropriate size room may not be available.
Professional skills	How will you use this learning?	I will display humility by being equally willing to present or sit in the mail room.
Personal learning	What did you learn?	I learned that I worry when others present my work and I am not there to correct them.
Personal learning	How did you learn it?	I learned this because I was sitting in the mail room while one of my team mates was presenting my work.
Personal learning	Why is this learning important?	This learning is important because it can happen again. Also, I could risk having to present my team mates' work.
Personal learning	How will you use this learning?	I will try to trust my team mates and make sure we always know each other's work and can present it to the satisfaction of all of us.

5.3   A BORING PROJECT: CHANGE OF TEAM RESPONSIBILITIES

LEARNING: POSITIVE ATTITUDE, DISCIPLINE, HUMILITY, LEADERSHIP SKILLS, AND EXPECTATION MANAGEMENT

We were on site at a major company. The project we worked on was very large and our part of it just small. We were entering answers from a questionnaire into a database. It was really boring and something somebody without an education could do. Not graduate students! We were quite upset about this also because the other groups seemed to have much more exciting projects than us. We discussed this with our instructor and were told that entering this data was as important as anything else and that we should practice having a positive attitude. We were also told that working in a company requires being willing to take all roles as requested. We were not happy about this.

Later, this project turned into a much more elaborate and exciting one because we were using all our data entries to deduce patterns of behavior and devise ways to change these behaviors. It turned out the boring work was extremely important, and we were entrusted with its proper content, the exciting down stream analysis, and giving suggestions for a future plan based on the data. We even developed creative and instructional materials for the employees, which were very exciting tasks, indeed.

REFLECTION

Level of learning	Question	Student answers
Professional skills	What did you learn?	I learned that I do not always appreciate all parts of a project and that all parts are important.
Professional skills	How did you learn it?	I learned this because I first only understood a very little part, the part I was working on (which I did not appreciate). Later, when the project evolved, I appreciated the former part.
Professional skills	Why is this learning important?	This learning is important because all parts of a project are essential.
Professional skills	How will you use this learning?	I will attempt to understand the project better and change my expectations so they are aligned with goals of the project.
Personal learning	What did you learn?	I learned that I sometimes expect too much of my part of the project and the visibility I want from it.
Personal learning	How did you learn it?	I learned this because initially my project seemed to be too easy (boring) and not important.
Personal learning	Why is this learning important?	This learning is important because I need to learn that I work on a project for the company's sake, not for my own sake. Also, I may not always understand how important a seemingly boring aspect may be or how my project will change.
Personal learning	How will you use this learning?	I will stay positive and display good self discipline and control when I am asked to do something seemingly boring. I trust that things will change and that exciting aspects will appear also.

5.4   OUT OF THE BOX: REQUIRING INNOVATION

LEARNING: TEAMWORK, AND CREATIVE THINKING

Our instructor told us we had to stretch ourselves. Create surprises for the company. Innovation. Not easy. How and what were the questions? This was something very unfamiliar to us. Some of the teams had boring and simple projects so they had to "add value" to their projects by expanding these and revising the project parameters, thereby demanding more of themselves. In fact, having a boring project turned out to be a real incentive for making these less boring because all the projects competed with each other in terms of content and innovation value. Other teams had enough to do just getting through all the data and analyzing it. These teams really had to manage their time and project well and much of their work had to be done on an individual basis. Yet other teams were focusing on regulations and it seemed impossible to create anything new there. But all teams were determined to do well, probably because innovation was a requirement.

After a while we got excited and "out of the box." The teams with "boring" projects had time to add new dimensions and creativity. They expanded the network of people to ask for advice, to provide new aspects, and to interact with (some as contractors, and some as clients). This was highly supportive for the teams who were busy analyzing data because they could piggy-back on many of the ideas. The teams doing regulatory work was also really inventive because they found ways to streamline the processes in a new and better fashion. We learned a lot about the incentives for innovation and how to make innovation part of our thought regimen.

REFLECTION

Level of learning	Question	Student answers
Professional skills	What did you learn?	I learned that innovation can be applied to any project.
Professional skills	How did you learn it?	I learned this because I was required to invent something in my project.
Professional skills	Why is this learning important?	An innovative mindset is of high value because of the potential for creating better solutions.
Professional skills	How will you use this learning?	I will always think of how I can improve the project and project products I am working on.
Personal learning	What did you learn?	I learned that being innovative is something I enjoy and also that it takes time away from simply performing a project.
Personal learning	How did you learn it?	I learned this because I had to finish a project while at the same time inventing something new and it was hard to do both well.
Personal learning	Why is this learning important?	This learning is important because I need to be a good manager of my time and also create new value for my future employer.
Personal learning	How will you use this learning?	I will attempt to plan my time well so I can enjoy thinking innovatively and at the same time get my work done as expected.

5.5   NERVOUS BUT NOT DEAD: CHANGE OF DELIVERY PERSON

LEARNING: COMMUNICATION SKILLS, AND DISCIPLINE

We were nearing the end of a project and all of us thought we were going to present our well-known part of the larger project. It turned out that was not the case. Instead one person would be assigned to present the entire project at the company for the industry professionals. This meant this person would also have to present the other groups' parts. All group leaders got a chance to try out for this. I did not think I was very good (I was nervous), but the instructor picked me as spokesperson for the entire class and the project. It was nerve wracking. I had never done this before and now my grade and everybody else's grades depended on how well I would do.

I worked diligently to come up with eloquent language, simple slides with essential information, and a nice flow to all of it. It took an enormous amount of practice. My classmates were a bit nervous on my behalf and we conducted a dress rehearsal. They were very helpful, albeit critical. I had a minor breakdown and blanked out. The instructor heard about it and called me to ask what was going on. The instructor also said that I was only asked to do this because of the absolute confidence that I would be able to do this well if I could overcome being nervous. I needed better self control. The next question was: "what is the worst that can happen?" I realized that I most likely would not die and that there are much more challenging situations in life. Then we practiced learning the first 5 minutes by heart. I was ready and much calmer. I gave a great talk the following day.

REFLECTION

Level of learning	Question	Student answers
Professional skills	What did you learn?	I learned that giving a good presentation requires mastering of good communication skills, practice and self control.
Professional skills	How did you learn it?	I learned this because I was required to present a project on behalf of the entire class.
Professional skills	Why is this learning important?	This learning is important because each presentation is a chance to either give a good or a bad impression and there are not many chances given.
Professional skills	How will you use this learning?	In the future I will plan and practice my talks diligently. I will make sure I know ahead of time what I want to say and the points I will make. I will also ask others to listen to the talk before the actual presentation takes place.
Personal learning	What did you learn?	I learned that I get very nervous before a presentation but also that I can survive giving one.
Personal learning	How did you learn it?	I learned this because I had to present the class project to industry professionals.
Personal learning	Why is this learning important?	This learning is important because I need to be calm so nervousness does not become a distraction for me or the audience.
Personal learning	How will you use this learning?	I will use this learning to practice self control when I have to give a presentation. I will remind myself that I most likely will survive to live another day.

5.6   WHAT ARE WE GOING TO DO: OPEN-ENDED PROJECTS

LEARNING: AMBIGUITY MANAGEMENT, AND INNOVATIVE THINKING

I used to be very good at giving exact and correct answers to questions. I still am, but how will I know when there is no exact solution to a problem and it is hard to know if a particular solution is the right one? This was the case in one of our studies. It was a huge challenge for me because I had to accept that there could be more than one solution to a problem.

We were given the task to optimize a production process. There are so many components to such a process and we had to determine what optimization meant exactly. The tricky part was it could be anything from creating more product, a better quality product, new innovative aspects of the product or the process, less mistakes in the process, etc.

For a while, we were paralyzed and overwhelmed because it seems like there was such an enormous lack of direction for us. Unfortunately, we still needed to finish our project by the given deadline (our grades depended on it) so we asked our instructor to point out the most important aspects. Our instructor was not sure and it seemed like importance was relative and dependent on many other factors.

After recovering from the unease, we decided to create a menu of recommendations for optimizations. Some related to innovation, some to quality control, and yet others to quantity. After reviewing our work, the instructor was able to pick two of our plans and give us direction for further and additional work. In fact, these suggestions caused the instructor to add entirely new challenges and aspects to our project. It seemed that possibly these two plans were equally good and that our work was part of a larger decision making process for the instructor.

REFLECTION

Level of learning	Question	Student answers
Professional skills	What did you learn?	I learned that sometimes it is not clear what the end-product of my work is neither for me nor for my superiors.
Professional skills	How did you learn it?	I learned this because my project was open-ended and required a set of creative solutions rather than one solution only.
Professional skills	Why is this learning important?	This learning is important because company work is highly dynamic and decisions can change along the way depending on new parameters and considerations.
Professional skills	How will you use this learning?	I will expect some uncertainty regarding direction and also be prepared to change my plans as needed.
Personal learning	What did you learn?	I learned that I feel uneasy if I do not know exactly what to do and where my project leads me.
Personal learning	How did you learn it?	I learned this because my project was full of ambiguity and required creative thinking and a high amount of flexibility.
Personal learning	Why is this learning important?	This learning is important because companies are places of innovation and if I am going to work in a company, I have to be able to work under these conditions.
Personal learning	How will you use this learning?	I will try to be excited about the open-ended nature of my projects. It is an opportunity for being creative in a need-based manner.

5.7   NOBODY IS AROUND: CHANGE IN INSTRUCTOR AVAILABILITY

LEARNING: FLEXIBILITY, TEAMWORK, AMBIGUITY MANAGEMENT, AND DISCIPLINE

All of us were deeply involved with our project at the company. Then we heard that our instructor's wife was having a baby and that he would not be available the next week. That was understandable, and we made sure we planned far ahead before he was gone. Later, we learned there had been some complications and that he was not going to return for a while. That was not good news. We had no idea how to continue the project at this point and who to ask for advice. Nobody else at the company seemed to know either. There was a sense of panic and also hopelessness. How could we finish the project without help? How would we know if the project was good enough?

We started talking to the other groups and instructors. We consulted new literature and took some risk in developing our own direction for the project. We invited other people to assist us and to critique our project. I think, we were more critical ourselves that we would have been if we had more help. We had to lead rather than to follow directions. In the end, we were able to produce a great result and felt a great deal of pride and ownership for what we were able to do. It was especially evident when we presented our work to the company executives. They were positively surprised about the direction of our work and our recommendations, which were heralded as truly upcoming and "out of the box."

REFLECTION

Level of learning	Question	Student answers
Professional skills	What did you learn?	I learned that being proactive in acquiring knowledge and alliances, and taking risk are great assets.
Professional skills	How did you learn it?	I learned this because we had to work independently and without an instructor for a while.
Professional skills	Why is this learning important?	This learning is important because I always want to perform at my very best.
Professional skills	How will you use this learning?	I choose to be proactive both when I have guidance and when I do not.
Personal learning	What did you learn?	I learned that it is hard (and a bit scary) for me to work without guidance but at the same time that it can be very rewarding.
Personal learning	How did you learn it?	I learned this because the instructor of my project became unavailable.
Personal learning	Why is this learning important?	This learning is important because instructors become unavailable at times, and also because I now know I can continue working well on the project myself.
Personal learning	How will you use this learning?	I will seek to always be able to understand the overall goals and direction of my project. I will practice being very independent and not afraid of leading the project if necessary.

5.8   MY IMPACT: ORAL TEAM MEMBER EVALUATION

LEARNING: POSITIVE ATTITUDE, CROSS-CULTURAL UNDERSTANDING, COMMUNICATION SKILLS, CRISIS MANAGEMENT, AND CONFLICT RESOLUTION

One of my team members was from another country. That is usually very exciting because cultures can vary so much. In this case, it was not beneficial for the team. This person was consistently coming across as extremely negative. The team was close to a crisis and our instructor intervened and said we needed to sit down and perform an oral evaluation under the instructor's guidance. We did. It was a great experience because the instructor was able to stay neutral and help us be assertive enough to voice our concerns. Each team member confronted the negative person and described which effect this negativity had on them. The funny thing was that the negative person had no clue that everybody had such a perception: "this is how we all communicate where I come from." Once specific examples were given and it was explained how these examples contributed negatively, it was possible to obtain commitment to a more positive style. Later, our negative team member really changed style and was thankful for the insight the evaluation had provided.

REFLECTION

Level of learning	Question	Student answers
Professional skills	What did you learn?	I learned that some people come across in negative ways without knowing it.
Professional skills	How did you learn it?	I learned this because I worked with such a person from another culture and this person was completely unaware of my perception.
Professional skills	Why is this learning important?	This learning is important because positive environments are most productive and perceived negativity may or may not be real, it could simply be a matter of style or misunderstanding of communication.
Professional skills	How will you use this learning?	In the future I will address perceived negativity immediately before it influences my project and the team morale.
Personal learning	What did you learn?	I learned that it is difficult for me to work in a team if there is a negative atmosphere. I also learned that oral evaluations are scary but can solve problems quickly.
Personal learning	How did you learn it?	I learned this because one of our team members seemed negative, and we were asked to perform an oral evaluation.
Personal learning	Why is this learning important?	This learning is important because in order for teams to work optimally, negativity must be minimized.
Personal learning	How will you use this learning?	I will be courageous and ask for feedback often from my team members about myself. I will also provide feedback to other team members, especially, if I feel somebody is creating an unproductive atmosphere.

CHAPTER 6

Measuring the Outcome

Professional skills are not easily measured, and this may in fact be a reason many professors want to avoid teaching such skills. Companies usually include assessment of professional skills during their formal performance evaluations. Typical areas of evaluation may include teamwork, punctuality, presentation skills, integrity, flexibility, organizational skills, and innovation. Evaluation of these areas may initially seem difficult or prone to subjectivity. This is usually countered by a broad scoring scale and scores may ultimately be summarized into a ranking status. It is of great importance to develop new instruments for assessment of professional skills so students and professors can perform this task with confidence.

6.1 PERFORMANCE LEVELS

Below (section 6.3) are some suggestions for rubric entries for a range of the essential professional skills described in this work and in Resources, reference [1]. Students can be scored or score them-selves according to their performance level (0, 1, 2, 3 or 4). This is done by asking how often a student conducts the tasks mentioned in each field of the rubric. Possible student answers are "never," "rarely," "often," and "always" at a given performance level. If a student answers "never" or "rarely" in a certain field the student is at a performance level below the level that is being addressed. The student should not proceed to the next field in the rubric but work on improving the skill set at the current level until the answer "often" or "always" can be given. When a student can answer "often" or "always" the student should continue to address the skill set at the following performance level. In other words: the two answers "never" and "rarely" prohibit students from proceeding to the next level score, whereas the answers "often" and "always" allows evaluating the performance at the follow-ing level. Similarly, instructor answers such as "not meeting expectations," "meeting expectations," and "exceeding expectations" can be used. The answer "not meeting expectations" does not permit further evaluation, whereas the answers "meeting expectations" and "exceeding expectations" create an opportunity to assess performance at the following level. The goal is to ensure that students are learning all essential professional skills, depicted in the rubric to a proficiency level of 4.

6.2   EXAMPLE OF SCORING PROFESSIONAL SKILLS PROFICIENCY: ORAL PRESENTATION

1. Possible answers: "never," "rarely," "often," and "always." Evaluation often performed as student self evaluation and/or evaluation by the instructor.

Level 0: Student does not master level 1 and answers "never" or "rarely" to the field: speaks with a clear and relaxed voice and begins by thanking the audience and the organizers.

Level 1: Student masters level 1 and answers "often" or "always" to the field: speaks with a clear and relaxed voice and begins by thanking the audience and the organizers. Student does not master level 2 and answers "never" or "rarely" to the field: presents message in an objective and clear manner. Student does not look at slides too often and preparation is apparent throughout presentation.

Level 2: Student masters level 1 and 2 and answers "often" or "always" to fields 1, and the additional field 2: presents message in an objective and clear manner. Student does not look at slides too often and preparation is apparent throughout presentation. Student does not master level 3 and answers "never" or "rarely" to the field: listens carefully to questions and makes sure they are fully understood before answering. Student is honest if his/her answer is unsure or unknown.

Level 3: Student masters level 1, 2, and 3 and answers "often" or "always" to fields 1, and 2, and the additional field 3: Listens carefully to questions and makes sure they are fully understood before answering. Student is honest if his/her answer is unsure or unknown. Student does not master level 4 and answers "never" or "rarely" to the field: shows sincerity and credibility before, during, and after presentation and is able to command the audience's attention.

Level 4: Student masters level 1, 2, 3, and 4 and answers "often" or "always" to the fields 1, 2, and 3, and the additional field 4: shows sincerity and credibility before, during, and after presentation and is able to command the audience's attention.

2. Possible answers: "not meeting expectations," "meeting expectations," and "exceeding expectations." Evaluation often performed by the instructor.

Level 0: Student does not master level 1 and instructor answers "not meeting expectations" to the field: speaks with a clear and relaxed voice and begins by thanking the audience and the organizers.

Level 1: Student masters level 1 and instructor answers "meeting expectations" or "exceeding expectations" to the field: speaks with a clear and relaxed voice and begins by thanking the audience and the organizers. Student does not master level 2 and instructor answers "not meeting expectations" to the field: presents message in an objective and clear manner. Student does not look at slides too often and preparation is apparent throughout presentation.

Level 2: Student masters level 1 and 2 and instructor answers "meeting expectations" or "exceeding expectations" to fields 1, and the additional field 2: presents message in an objective and clear manner. Student does not look at slides too often and preparation is apparent throughout presentation. Student does not master level 3 and instructor answers "not meeting expectations" to the field: listens carefully to questions and makes sure they are fully understood before answering. Student is honest if his/her answer is unsure or unknown.

Level 3: Student masters level 1, 2, and 3 and instructor answers "meeting expectations" or "exceeding expectations" to fields 1, and 2, and the additional field 3: listens carefully to questions and makes sure they are fully understood before answering. Student is honest if his/her answer is unsure or unknown. Student does not master level 4 and instructor answers "not meeting expectations" to the field: shows sincerity and credibility before, during, and after presentation and is able to command the audience's attention.

Level 4: Student masters level 1, 2, 3, and 4 and instructor answers "meeting expectations" or "exceeding expectations" to the fields 1, 2, and 3, and the additional field 4: shows sincerity and credibility before, during, and after presentation and is able to command the audience's attention.

Students should be subjected to the same evaluation regularly in order to track student growth. Professors can devise their own system of extrapolation from a compilation of the scores to a particular grade. These may vary depending on the length of exposure to learning and the learning goals set forth during the exposure period. As an example, students may get a top score (A) if they reach level 1 in their first semester, level 2 in their second semester, etc. In addition, this score may be further defined by how well the student masters the skills. Students who can answer "always" concomitantly with an instructor evaluation of "exceeding expectations" should earn a higher grade (A+) than students who can answer "often" with an instructor evaluation of "meeting expectations." Transparency regarding the proportional influence of the professional skills evaluation on the final grade is an important aspect in grading of these assets.

6.3   PROJECT CONTEXT AND DETAILS

1	2	3	4
Describes mission for the project	Describes mission for the project	Describes mission for the project	Describes mission for the project
	Describes specific goals of the project, identifies stakeholders and impact of the project	Describes specific goals of the project, identifies stakeholders and impact of the project	Describes specific goals of the project, identifies stakeholders and impact of the project
		Develops a project plan and time table for deliverables, follows project plan, meets deadlines, and produces acceptable deliverables	Develops a project plan and time table for deliverables, follows project plan, meets deadlines, and produces acceptable deliverables
			Uses project as opportunity to foster professional network and skills

6.4   PLANNING AND DISCIPLINE

1	2	3	4
Consistently arrives on time	Consistently arrives on time	Consistently arrives on time	Consistently arrives on time
	Thinks and plans out work before beginning tasks	Thinks and plans out work before beginning tasks	Thinks and plans out work before beginning tasks
		Understands exact deliverables and produces them in a timely fashion. Is truthful, reports errors and takes responsibility	Understands exact deliverables and produces them in a timely fashion. Is truthful, reports errors and takes responsibility
			Perseveres in accomplishing tasks under adverse conditions

6.5   CHANGE MANAGEMENT

1	2	3	4
Accepts change in the work environment and in expectations	Accepts change in the work environment and in expectations	Accepts change in the work environment and in expectations	Accepts change in the work environment and in expectations
	Understands triggers of change and how the change may cause a revision in priorities and possible reorganization	Understands triggers of change and how the change may cause a revision in priorities and possible reorganization	Understands triggers of change and how the change may cause a revision in priorities and possible reorganization
		Adapts to change by revising deliverables and priorities and helps others to adapt	Adapts to change by revising deliverables and priorities and helps others to adapt
			Knows where one's work fits into the priority list. Invites change as something positive that can foster improvement and innovation

6.6   INNOVATIVE THINKING

1	2	3	4
Sees innovative thinking as important for progress	Sees innovative thinking as important for progress	Sees innovative thinking as important for progress	Sees innovative thinking as important for progress
	Keeps up to date with new technology, equipment, literature, in fields of relevance	Keeps up to date with new technology, equipment, literature, in fields of relevance	Keeps up to date with new technology, equipment, literature, in fields of relevance
		Studies other fields and cultures in search of ideas that can spur innovation	Studies other fields and cultures in search of ideas that can spur innovation
			Embraces risk and creates new ways to solve problems

6.7 AMBIGUITY MANAGEMENT

1	2	3	4
Accepts ambiguity and that work conditions are dynamic	Accepts ambiguity and that work conditions are dynamic	Accepts ambiguity and that work conditions are dynamic	Accepts ambiguity and that work conditions are dynamic
	Adapts and helps others adapt to situations of ambiguity	Adapts and helps others adapt to situations of ambiguity	Adapts and helps others adapt to situations of ambiguity
		Identifies the main reasons of ambiguity and actively seeks to decrease these	Identifies the main reasons of ambiguity and actively seeks to decrease these
			Suggests ways to use ambiguity to create the best possible decision

6.8   TEAMWORK AND LEADERSHIP

1	2	3	4
Takes responsibility for own and team's deliverables	Takes responsibility for own and team's deliverables	Takes responsibility for own and team's deliverables	Takes responsibility for own and team's deliverables
	Meets with team members regularly, sets agendas, distributes tasks, understands the strengths and weaknesses of the team members	Meets with team members regularly, sets agendas, distributes tasks, understands the strengths and weaknesses of the team members	Meets with team members regularly, sets agendas, distributes tasks, understands the strengths and weaknesses of the team members
		Takes time to actively listen and create a positive work environment, solves conflicts quickly and assertively	Takes time to actively listen and create a positive work environment, solves conflicts quickly and assertively
			Shows vision, enthusiasm, perseverance. Displays personal humility: gives credit to team, takes blame for mistakes

6.9   ORAL PRESENTATION

1	2	3	4
Speaks with a clear and relaxed voice and begins by thanking the audience and the organizers	Speaks with a clear and relaxed voice and begins by thanking the audience and the organizers	Speaks with a clear and relaxed voice and begins by thanking the audience and the organizers	Speaks with a clear and relaxed voice and begins by thanking the audience and the organizers
	Presents message in an objective and clear manner. Does not look at slides too often and preparation is apparent throughout presentation	Presents message in an objective and clear manner. Does not look at slides too often and preparation is apparent throughout presentation	Presents message in an objective and clear manner. Does not look at slides too often and preparation is apparent throughout presentation
		Listens carefully to questions and makes sure they are fully understood before answering. Is honest if an answer is unsure or unknown	Listens carefully to questions and makes sure they are fully understood before answering. Is honest if an answer is unsure or unknown
			Shows sincerity and credibility before, during, and after presentation and is able to command the audience's attention

6.10   WRITTEN PRESENTATION

1	2	3	4
Writes work that is complete, grammatically correct, and well organized	Writes work that is complete, grammatically correct, and well organized	Writes work that is complete, grammatically correct, and well organized	Writes work that is complete, grammatically correct, and well organized
	Places content in the appropriate sections without repetition. Provides a clear, objective and respectful message	Places content in the appropriate sections without repetition. Provides a clear, objective and respectful message	Places content in the appropriate sections without repetition. Provides a clear, objective and respectful message
		Displays good transitions between the various sections of the work. Can justify the inclusion of each sentence	Displays good transitions between the various sections of the work. Can justify the inclusion of each sentence
			Ensures that work contains implications for the future and when applicable, extrapolations, not exaggerations

6.11   INTERPERSONAL COMMUNICATION

1	2	3	4
Knows own communication style. Is knowledgeable about assertive communication style	Knows own communication style. Is knowledgeable about assertive communication style	Knows own communication style. Is knowledgeable about assertive communication style	Knows own communication style. Is knowledgeable about assertive communication style
	Uses assertive communication to express own wishes and takes responsibility for own feelings	Uses assertive communication to express own wishes and takes responsibility for own feelings	Uses assertive communication to express own wishes and takes responsibility for own feelings
		Understands that listening is important for good communication and knows the concept of active listening	Understands that listening is important for good communication and knows the concept of active listening
			Uses active listening to make sure the speaker's message and feelings are understood and that the message is perceived as intended

6.12 ATTITUDE

1	2	3	4
Respects other people's opinions and accepts that everyone can improve	Respects other people's opinions and accepts that everyone can improve	Respects other people's opinions and accepts that everyone can improve	Respects other people's opinions and accepts that everyone can improve
	Displays a positive attitude, including at times when improvements are requested	Displays a positive attitude, including at times when improvements are requested	Displays a positive attitude, including at times when improvements are requested
		Attempts to incorporate other people's suggestions for improvement	Attempts to incorporate other people's suggestions for improvement
			Encourages others to be positive, leading by example through being open for self-improvement

6.13   CONFLICT RESOLUTION

1	2	3	4
Sees conflict as a creative opportunity	Sees conflict as a creative opportunity	Sees conflict as a creative opportunity	Sees conflict as a creative opportunity
	Uses assertive communication and active listening skills to identify cause for conflict	Uses assertive communication and active listening skills to identify cause for conflict	Uses assertive communication and active listening skills to identify cause for conflict
		Collaborates with conflicting parties to formulate a plan for resolution	Collaborates with conflicting parties to formulate a plan for resolution
			Obtains commitment from conflicting parties to follow plan. Successfully manages the conflict

6.14   CROSS-CULTURAL UNDERSTANDING

1	2	3	4
Understands and respects that people from other cultures may have different work ethics and value systems	Understands and respects that people from other cultures may have different work ethics and value systems	Understands and respects that people from other cultures may have different work ethics and value systems	Understands and respects that people from other cultures may have different work ethics and value systems
	Is able to view own culture and value system from another culture's point of view	Is able to view own culture and value system from another culture's point of view	Is able to view own culture and value system from another culture's point of view
		Manages cultural conflicts and works to find the optimal solution to problems	Manages cultural conflicts and works to find the optimal solution to problems
			Views differences as an opportunity to learn, grow, and improve on best practices

6.15   KNOWLEDGE ABOUT "SELF"

1	2	3	4
Knows and accepts own strengths and weaknesses	Knows and accepts own strengths and weaknesses	Knows and accepts own strengths and weaknesses	Knows and accepts own strengths and weaknesses
	Accepts criticism of self with confidence; sees criticism as opportunity to improve	Accepts criticism of self with confidence; sees criticism as opportunity to improve	Accepts criticism of self with confidence; sees criticism as opportunity to improve
		Works to improve strengths and strengthen areas of weakness	Works to improve strengths and strengthen areas of weakness
			Adjusts and performs personal improvement according to needs of the work environment

6.16   EXPECTATION MANAGEMENT

1	2	3	4
Knows what is expected in terms of performance and goals	Knows what is expected in terms of performance and goals	Knows what is expected in terms of performance and goals	Knows what is expected in terms of performance and goals
	Regularly completes self assessment to make sure one is on track	Regularly completes self assessment to make sure one is on track	Regularly completes self assessment to make sure one is on track
		Sees performance reviews as an opportunity to revise goals and expectations	Sees performance reviews as an opportunity to revise goals and expectations
			Receives feedback in a positive manner, keeping a professional demeanor, and is willing to improve and learn new skills

Bibliography

[1] Borbye, L.: *Secrets to Success in Industry Careers: Essential Skills for Science and Business.* 224 pp. Elsevier Science (Academic Press), London, United Kingdom (2008). 4, 9, 47

[2] Borbye, L., Stocum, M., Woodall, A., Pearce, C., Sale, E., Barrett, W., Clontz, L., Peterson, A. and Shaeffer, J.: *Industry Immersion Learning: Real-life Case Studies in Biotechnology and Business.* 205 pp. Wiley Blackwell, Weinheim, Germany (2009). 9, 11

Author's Biography

LISBETH BORBYE

Lisbeth Borbye, Ph.D., is a pioneer in alliance building between industry and universities, and development of innovative, need-based instructional materials for students and their professors. Dr. Borbye is Assistant Dean for Professional Education at North Carolina State University and Director of the University of North Carolina's System-wide Professional Science Master's Initiative. Professional science Master's programs seek to meet the needs for an improved graduate workforce in industry, non-profit organizations, and government by providing interdisciplinary career-specific education. Dr. Borbye has been instrumental in establishing and directing the highly successful professional science Master's program called Microbial Biotechnology at North Carolina State University. The program is known for its elaborate interaction with industry partners in the Research Triangle Park, its dual degree option, and its successful alumni. Dr. Borbye is the author of the book entitled "Secrets to Success in Industry Careers: Essential Skills for Science and Business," (Academic Press) and a co-author of "Industry Immersion Learning: Real-Life Case Studies in Biotechnology and Business" (Wiley-Blackwell). Dr. Borbye is also a co-Founder of the National Professional Science Master's Association.

Printed in the United States
by Baker & Taylor Publisher Services